To Jonah, the biggest fan of fans.

Copyright © 2021 by Bethany Tabieros

All rights reserved. No part of this book may be reproduced or used in any manner without written permission of the copyright owner.

ISBN-13: 9798588249195

Celebrating Fans

by Bethany Tabieros

Ceiling fans are everywhere

Have you ever noticed how often you see ceiling fans? You find them in homes, schools and businesses. 75% of American homes have ceiling fans.

Fans have been in use for more than 100 years. They keep us cool and look beautiful too. Let's learn more about these spinning wonders.

Keeping us cool

When a fan is turned on, the blades push the air across your skin. The feeling of air moving along your skin helps to keep you cool.

Fans don't cool the air like air conditioning. They only make the air *feel* cooler to you. If you are not in the room, there is no purpose to having the fan turned on.

Blades push the air

You can find fans that have as little as 2 blades and some have many more. The more blades, the more air that is moved. More air movement will make you feel cooler.

The longer the blades, the more air it can push and the further the air moves. Larger rooms need fans with longer blades to move air throughout the room.

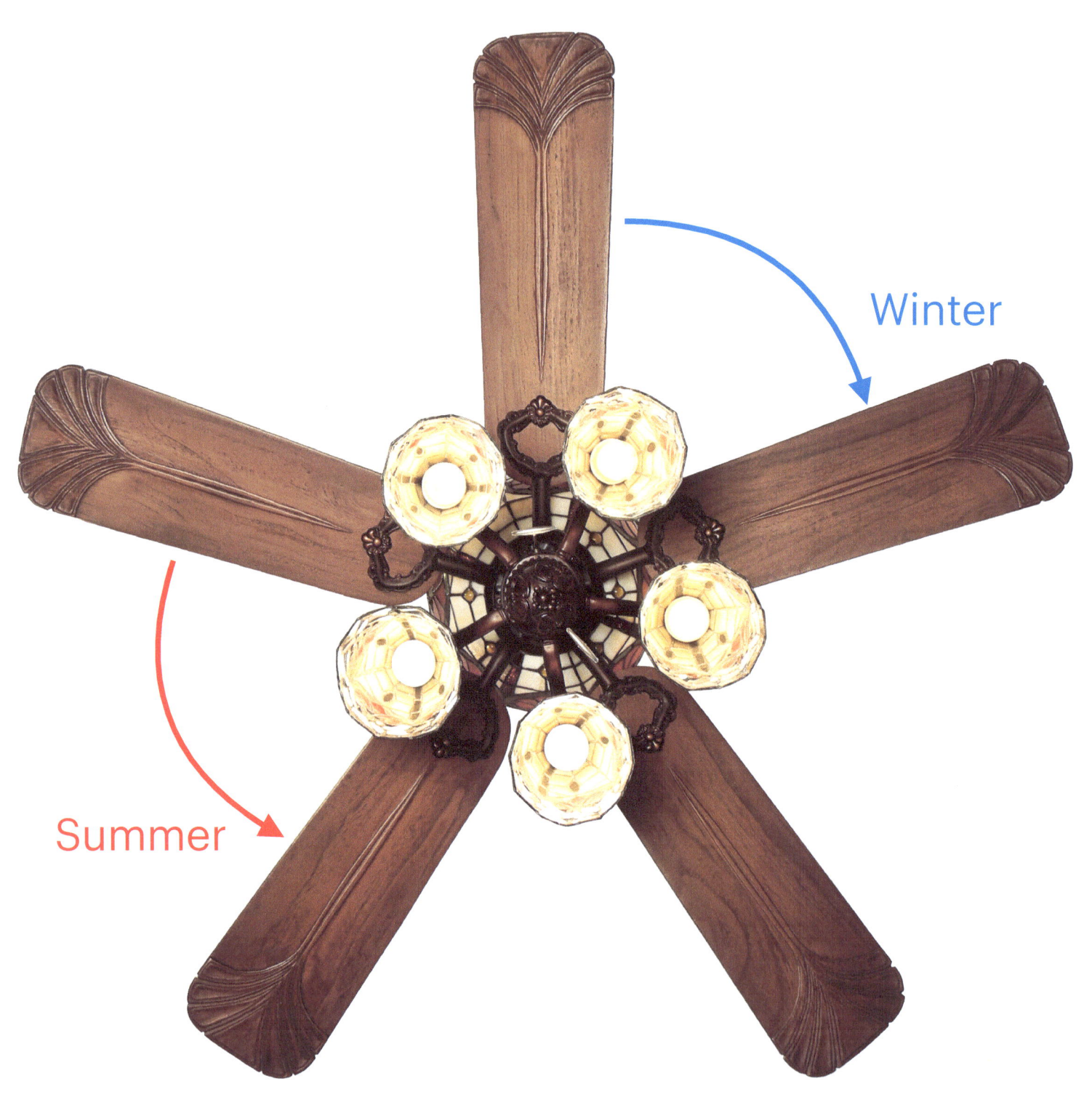

Summer and Winter

Fans can be used all year round. Many fans have a switch to change the direction the blades spin to either push air downward or pull air upward.

In *Summer,* the blades should spin *counterclockwise* to push air downward and help you feel cooler.

In *Winter,* changing the spin to *clockwise* will pull cold air upward and push warm air down so that you feel warmer.

Ceiling fans come in all shapes and sizes

The basic fan usually has a few blades and many have a light, but there are so many variations to fit different rooms, styles and purposes.

Different Designs

There are so many different ways a fan can look. This fan will keep you cool and looks pretty cool too!

Changing the direction the blades face affects where the air moves. This one helps push air into the corners of the room.

Scavenger hunt

Take a look around. See how many different kinds of ceiling fans you can find.

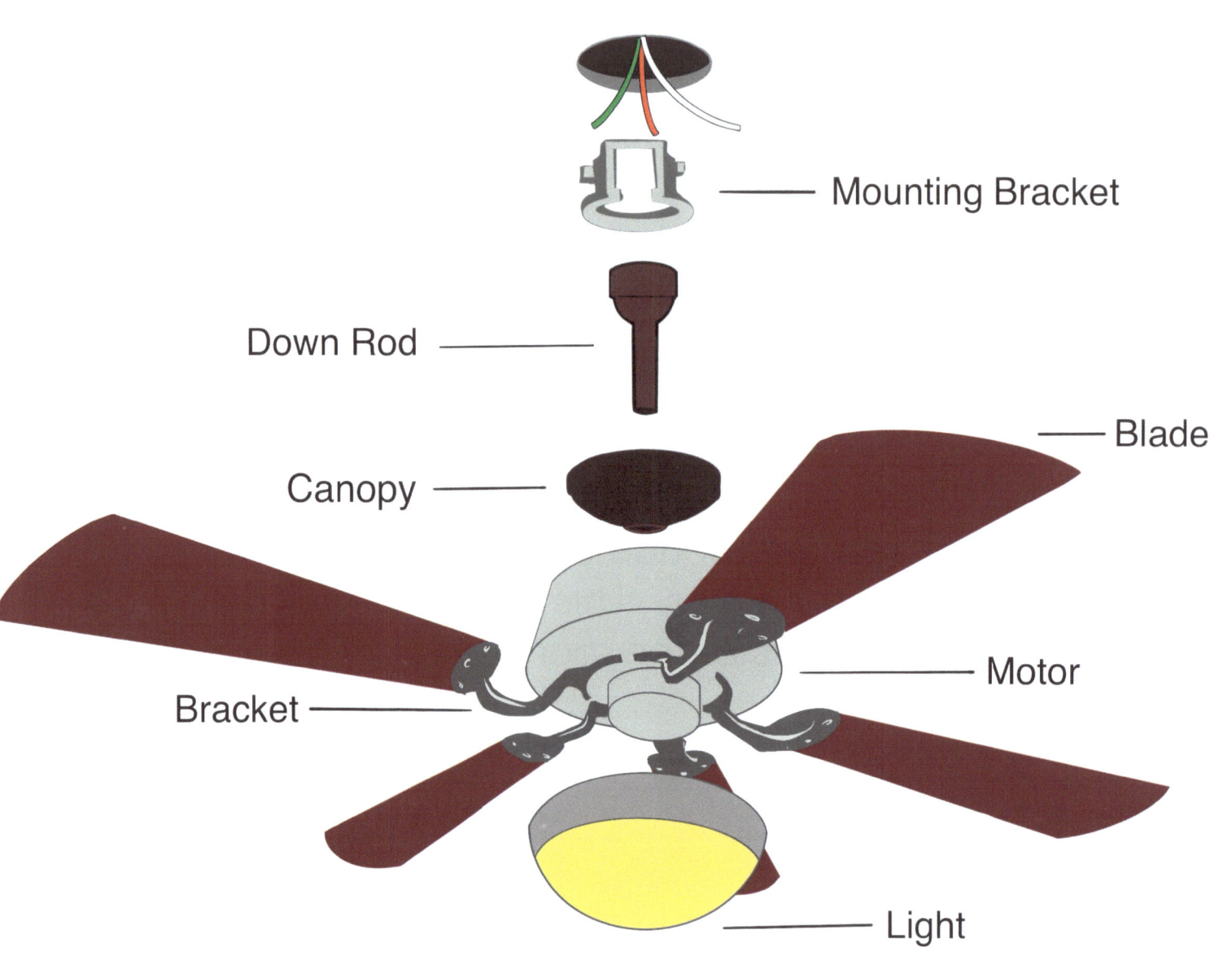
Illustration by Dennis Tabieros

Engineering marvels

Let's take a look at the parts that make fans work. Fans have wires that connect into the ceiling to provide electricity and connect to a chain, switch or remote control.

The down rod comes in different lengths to bring the fan closer to the ceiling or lower in the room. Blades are attached to the motor, which helps them move.

Beautiful styles

Fans come in all different shapes and sizes. They are as diverse as their owners. The blades can be customized to match your personal style. These blades look like leaves.

Reach for the rainbow

Fans can come in all types of colors. Different colors and designs add beauty and fun to the room. This fan looks like the sun!

Scavenger Hunt

Have you ever seen a fan with different colors? Keep an eye out to see how many different colors you can find.

Fun with lights

While some fans have no light at all, some have multiple lights. You can find lamps to match all sorts of styles. Look at the beautiful stained glass lights here.

This fan has multiple beautiful stained glass lights. Can you count them all?

Ceiling elegance

This fan has a chandelier in place of standard light. Multiple small prisms refract the light to make a beautiful display.

The right fan can bring beauty and elegance to any space.

Now it's your turn.

What kind of fan would you design? Would it be fun? Colorful? Crazy?

Let your imagination run WILD and maybe you will create the next ceiling fan inspiration!

To all the budding fan engineers out there - let yourself be inspired by the next ceiling fan you see. I can't wait to see what new designs you invent!

To my husband, Dennis, thank you for encouraging me to write this book, lending your artistic talent and always keeping me on my toes. To my kiddos, Gideon and Jonah, thank you for inspiring me every day.

Special thanks to Fanzart Fans and Tarun Lala for sharing their enthusiasm and allowing me to share pictures of their beautiful ceiling fans. Check out fanzartfans.com to see more gorgeous fans.

www.ingramcontent.com/pod-product-compliance
Lightning Source LLC
Chambersburg PA
CBHW051824210526
45473CB00005B/1734